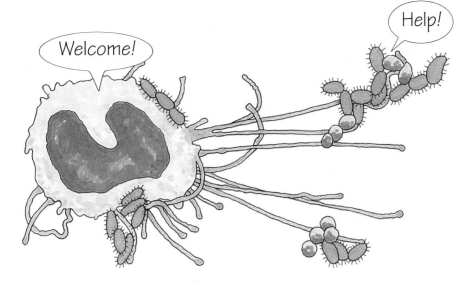

GERM ZAPPERS

Fran Balkwill & Mic Rolph

Cold Spring Harbor Laboratory Press

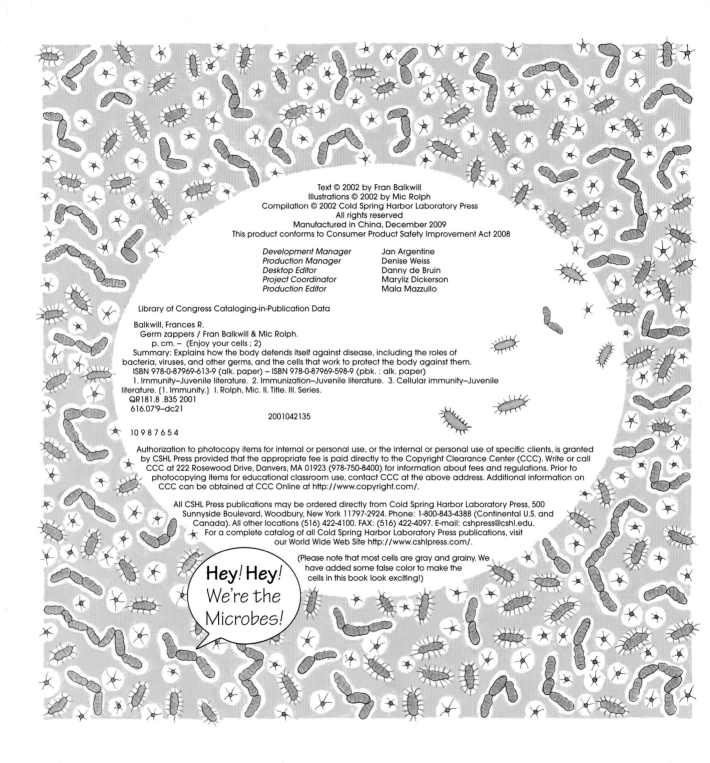

Development Manager	Jan Argentine
Production Manager	Denise Weiss
Desktop Editor	Danny de Bruin
Project Coordinator	Maryliz Dickerson
Production Editor	Mala Mazzullo

Library of Congress Cataloging-in-Publication Data

Balkwill, Frances R.
  Germ zappers / Fran Balkwill & Mic Rolph.
    p. cm. – (Enjoy your cells ; 2)
  Summary: Explains how the body defends itself against disease, including the roles of
bacteria, viruses, and other germs, and the cells that work to protect the body against them.
  ISBN 978-0-87969-613-9 (alk. paper) – ISBN 978-0-87969-598-9 (pbk. : alk. paper)
  1. Immunity–Juvenile literature. 2. Immunization–Juvenile literature. 3. Cellular immunity–Juvenile
literature. (1. Immunity.) I. Rolph, Mic. II. Title. III. Series.
  QR181.8 .B35 2001
  616.07'9–dc21
                                    2001042135

10 9 8 7 6 5 4

Authorization to photocopy items for internal or personal use, or the internal or personal use of specific clients, is granted
by CSHL Press provided that the appropriate fee is paid directly to the Copyright Clearance Center (CCC). Write or call
CCC at 222 Rosewood Drive, Danvers, MA 01923 (978-750-8400) for information about fees and regulations. Prior to
photocopying items for educational classroom use, contact CCC at the above address. Additional information on
CCC can be obtained at CCC Online at http://www.copyright.com/.

All CSHL Press publications may be ordered directly from Cold Spring Harbor Laboratory Press, 500
Sunnyside Boulevard, Woodbury, New York 11797-2924. Phone: 1-800-843-4388 (Continental U.S. and
Canada). All other locations (516) 422-4100. FAX: (516) 422-4097. E-mail: cshpress@cshl.edu.
For a complete catalog of all Cold Spring Harbor Laboratory Press publications, visit
our World Wide Web Site http://www.cshlpress.com/.

Hey! Hey!
We're the
Microbes!

(Please note that most cells are gray and grainy. We
have added some false color to make the
cells in this book look exciting!)

Planet Earth can be a dangerous place for all living creatures, including you. You can usually escape from erupting volcanoes and floods.

You can protect your body from the blazing Sun and freezing snow. But wherever you live and whatever the weather, you cannot escape **GERMS!**

Your body is made of millions and millions of complicated cells. Some of the most important cells protect you from invisible germs and attack any that dare to invade your body.

Well, we didn't know that!

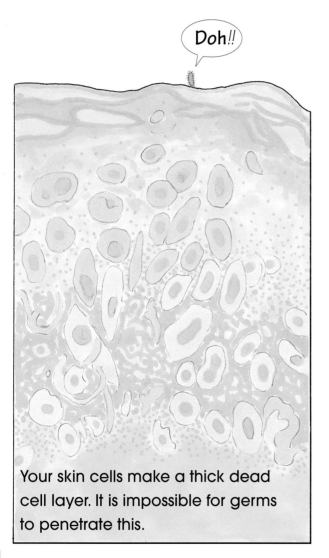

Doh!!

Your skin cells make a thick dead cell layer. It is impossible for germs to penetrate this.

Cells up your nose and down your lungs are fringed with microscopic hairs that waft away dirt, dust, and germs. They also make sticky mucus (mew-cuss) to trap germs.

Stomach cells make a deadly acid that destroys anything suspicious.

Tear duct cells make a germ zapping chemical that bathes your eyes all day and all night long.

But sometimes and somehow, germs still get through. Then they have to deal with the most cunning defender cells of all—

GERM ZAPPERS!

5

In the front line are germ-greedy cells called neutrophils (new-tro-fils). They are not picky, they attack and devour anything dangerous.

Neutrophils patrol your entire blood stream. When they sense trouble they stick to the walls of the blood vessel nearest the danger zone. They quickly squeeze their way between the blood vessel cells and crawl toward the trouble.

What a squeeze!

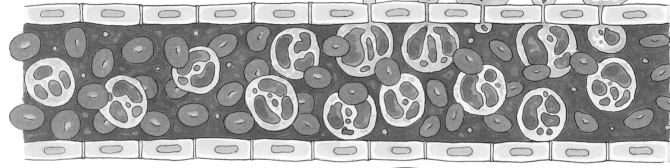

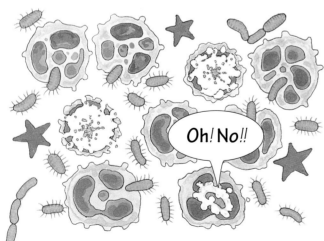

Oh! No!!

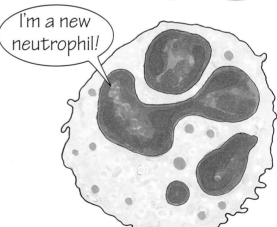

I'm a new neutrophil!

Neutrophils love to dine on juicy germs. Each cell is full of powerful chemical weapons that zap aliens. This is a deadly mission—the neutrophil dies as well!

Every hour of the day, stem cells in your bone marrow make about five thousand million new neutrophils.

Another cell that prowls through your body is the "Natural Killer" cell.
It hunts for cells that have been infected by germs.

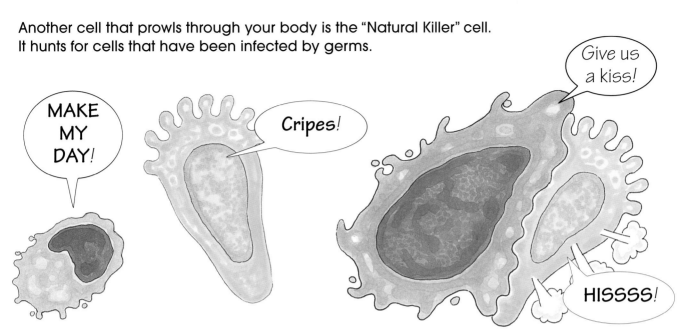

Infected cells can be dangerous—but the Natural Killer cell knows exactly what to do!...

...It sneaks up close and punches tiny holes in the membrane of the infected cell.

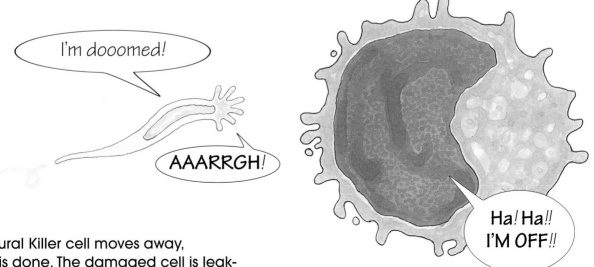

The Natural Killer cell moves away, the job is done. The damaged cell is leaking; soon it shrivels and dies.

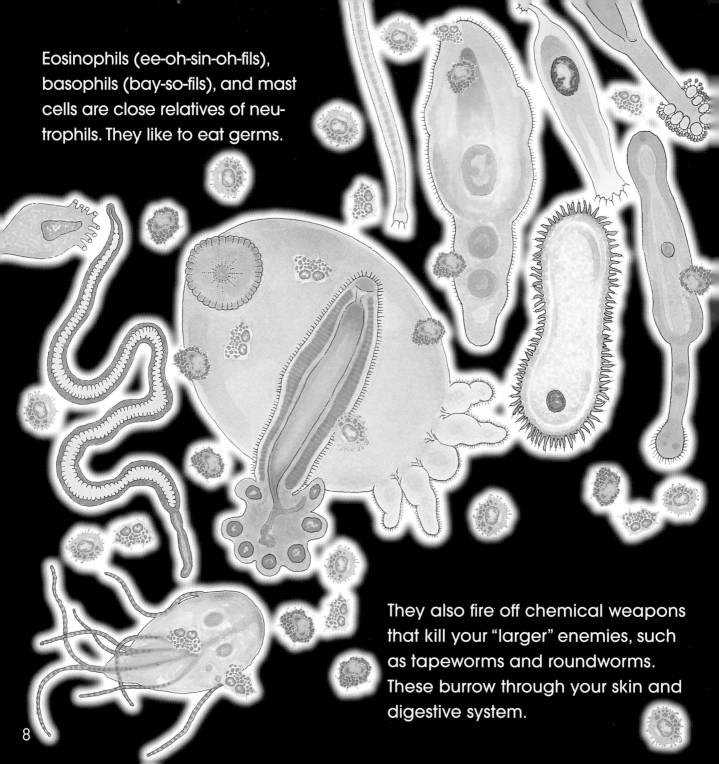

Eosinophils (ee-oh-sin-oh-fils), basophils (bay-so-fils), and mast cells are close relatives of neutrophils. They like to eat germs.

They also fire off chemical weapons that kill your "larger" enemies, such as tapeworms and roundworms. These burrow through your skin and digestive system.

Macrophages get rid of your rubbish!

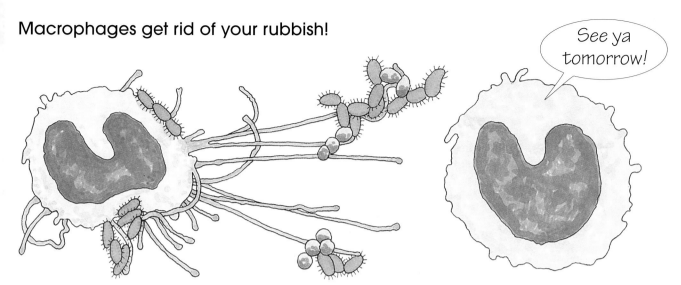

They clear up the mess whenever you are ill or injured and wherever germs and dirt collect.

Macrophages do not usually expire when they meet a germ. They live to fight another day!

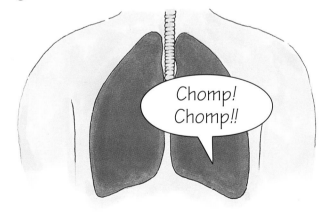

Every breath you take is full of microscopic dust and dirt. Millions of macrophages gobble up this junk to keep your lungs clean and tidy.

All these attacking cells belong to your innate (in-ate) immune (im-mewn) system. They react to invader dangers and they do it fast! Normally, they are all you need to keep you fit and healthy.

Sometimes, skulking germs manage to outwit these germ zappers. That is when you need your adaptive (add-ap-tiv) immune system.
And the cell that gets this going is called a dendritic (den-drit-ick) cell.

This is a highly efficient guard cell. Wherever the outside world meets your insides, you will find dendritic cells waiting patiently, ever watchful...

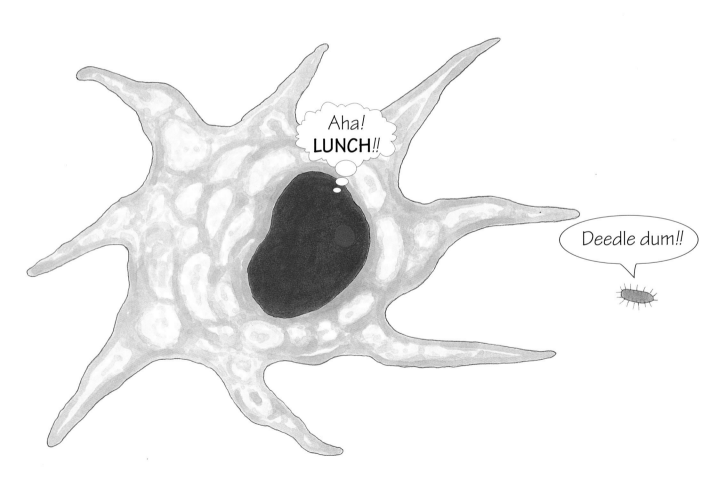

Dendritic cells capture germs, then zoom off to one of the best meeting places in your body for germ zappers. It is called a lymph (limf) node. In lymph nodes, dendritic cells alert the most sophisticated germ zapping cells of all, lymphocytes (lim-foe-sites).

Strange though it may seem, you have thousands of different lymphocyte squads, each one programmed to fight a particular type of germ. When a patrolling squad finds a dendritic cell that has captured its own germ, the lymphocytes multiply into millions and millions—armed to fight that particular germ and no other.

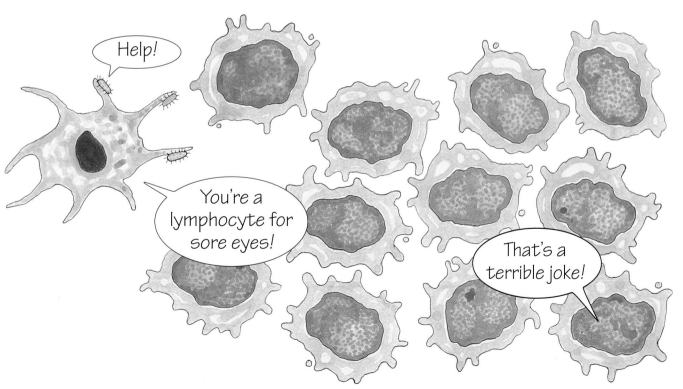

Some lymphocytes attack and kill every cell in the body that is infected with their particular type of germ.

Other lymphocytes release antibodies (an-tee-bod-ees). Antibodies are "guided missiles" that cling to "their" germ.

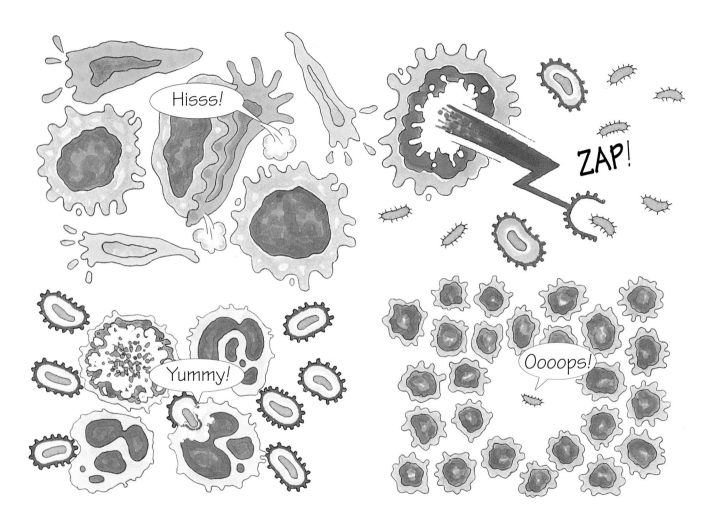

An antibody-coated germ makes a really tasty meal for macrophages and neutrophils. An antibody-coated germ is no longer harmful.

And if that germ ever dares to invade your body again, the lymphocyte squad will recognize it and attack, in double-quick time!

Your germ zapper defender cells need to be able to contact each other really quickly, if they are to work to keep you healthy. Defender cells send each other messages using special chemicals called cytokines (sigh-toe-kines).

Cytokines summon more germ zapping cells to a trouble spot and make them absolutely ravenous for the enemy.

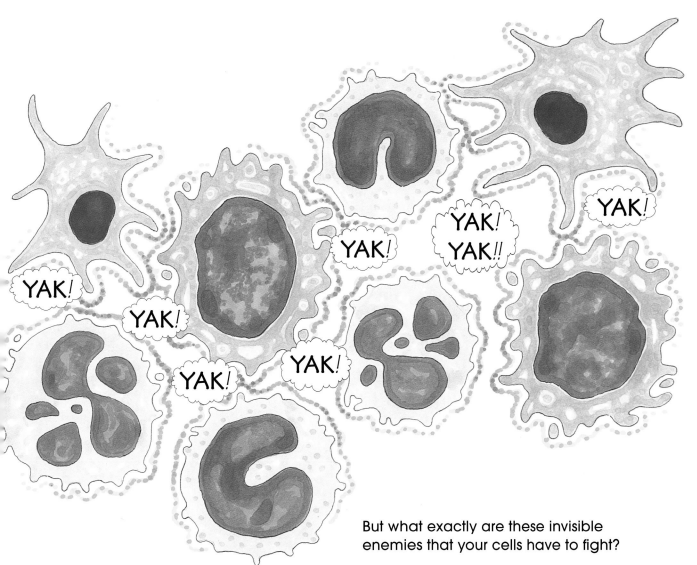

But what exactly are these invisible enemies that your cells have to fight?

BACTERIA!

Most bacteria are your friends and allies, not enemies. They are microscopic single-cell creatures, about one hundred times smaller than any human cell. They live almost everywhere on Planet Earth, from the frozen Antarctic, to boiling hot springs. They are found in the soil, where they support the fragile web of life.

There are even bacteria living inside *you.* They digest the food that you cannot and make vitamin K, which helps your blood to clot.

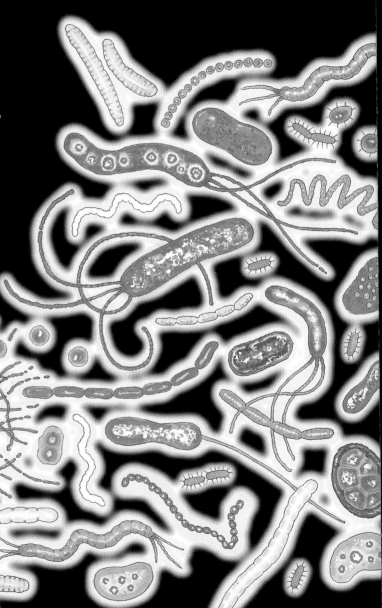

Only a few types of bacteria can make you ill. These menacing microbes multiply inside your body, or in the food that you eat.

As they grow, they release dangerous poisons that can make you feel very sick.

As usual, your neutrophil cavalry will seek out and destroy most invading bacteria.

But sometimes there are just too many, and the neutrophils are overwhelmed.

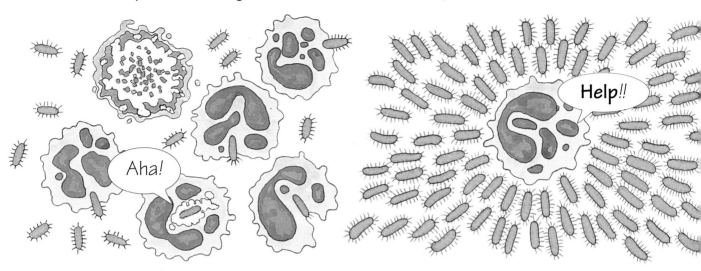

Dendritic cells are next on the case! They are busy alerting lymphocyte squads in the nearest lymph nodes.

The lymphocyte squad multiplies and makes antibody weapons that attack the dangerous bacteria.

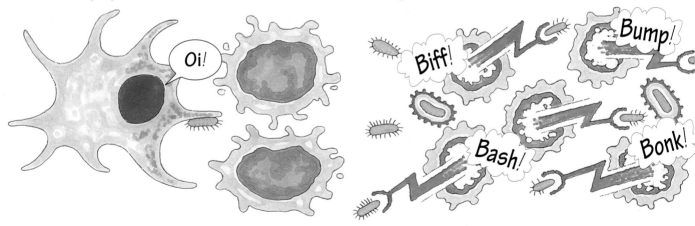

Most of the time, your germ zapping cells win these battles, but sometimes they need help. Your doctor may give you special medicines called antibiotics (an-tee-by-ot-icks), which kill bacteria, but not human cells.

16

One hundred years ago, many millions of children died each year of infections caused by bacteria. The children were often poor, living in cramped and dirty homes. Invisible microbes easily polluted their food and water.

Scientists and doctors in the 21st century understand a lot about bacteria. They have taught us how to keep food and homes safe from this hidden micro-menace. But in some parts of the world, where poverty, war, and famine still exist, young children continue to die from bacterial infections.

And then there are even smaller and
more loathsome enemies—

VIRUSES!

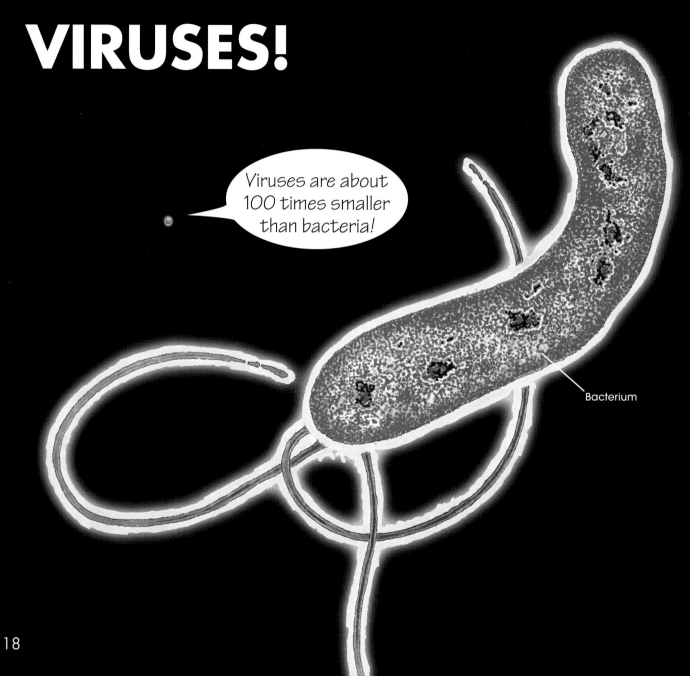

Viruses are about
100 times smaller
than bacteria!

Bacterium

Almost everyone (including you!) will suffer from a virus infection this year. But although viruses can make you feel quite ill, your germ zappers can beat most of them and you will soon feel better.

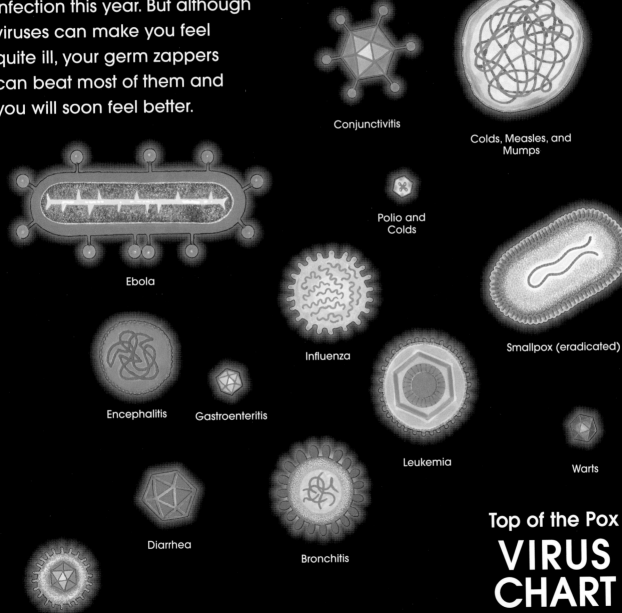

Conjunctivitis

Colds, Measles, and Mumps

Polio and Colds

Ebola

Influenza

Smallpox (eradicated)

Encephalitis

Gastroenteritis

Leukemia

Warts

Diarrhea

Bronchitis

Top of the Pox

VIRUS CHART

Rubella

Viruses are devious. These tiny alien invaders, about 100 times smaller than bacteria, cannot do their deadly work without cells in your body.

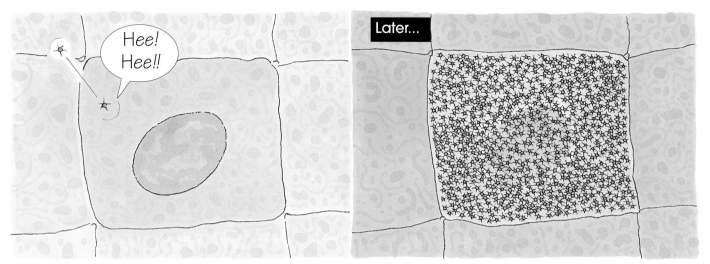

Viruses cause trouble by getting inside your body and then inside your cells...

...Once inside a cell, the viruses take over. They turn the cell into a production line making many more viruses.

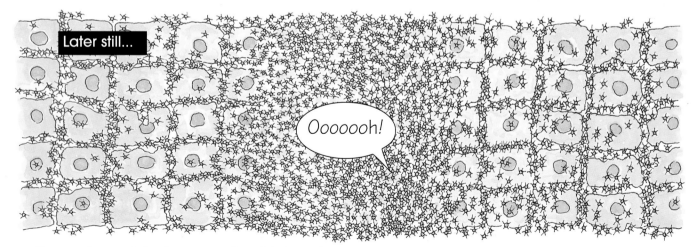

These viruses infect more cells, which make more viruses, which infect more cells, and so on until you feel really ill!

Your friend in the back row propels billions of virus particles across the school bus...

...Virus-filled droplets are sucked down inside your body when you breathe.

The virus makes an enzyme that chomps through the protecting mucus. It just takes one single virus to get inside one single cell beneath the mucus to start you feeling really sick.

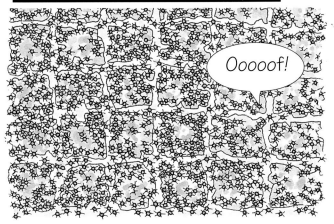

You feel OK, but the virus is taking over. Each infected cell makes ten thousand or more viruses each day.

Your Natural Killer cells are overwhelmed. But dendritic cells alert the mighty lymphocyte squads.

Cells in your lungs, breathing tubes and nose become virus factories —then they die!

Emergency! Cytokine messages call defender cells to battle. Hot, shivery and aching, you cough, sneeze and wheeze. Guess what? **You have the flu!**

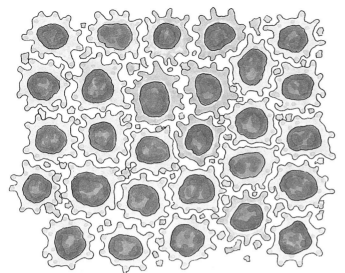

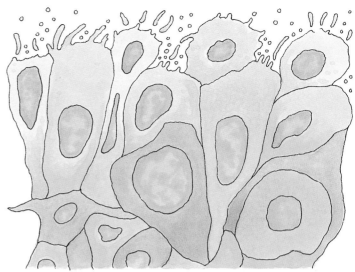

At last there are enough lymphocytes to destroy infected cells—and the virus does not like the heat of your fever...

...New lining cells replace those that die, and macrophages and dendritic cells return to their lairs.

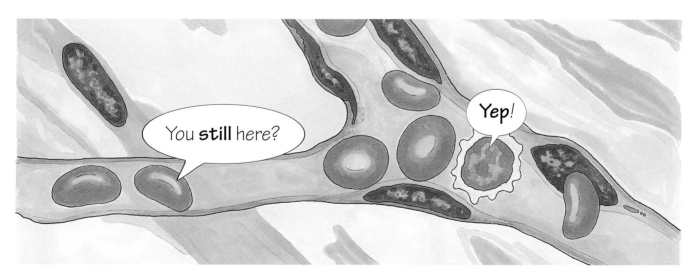

Powerful antibody missiles seek out the viruses outside your cells and remain in your bloodstream for several months. Leftover lymphocytes die off—only a few remain to patrol the body in case that virus attacks again.

Life would be very difficult without your germ zapper cells! We know this because there are some children whose defender cells do not work correctly.

A few kids are born with neutrophils that cannot destroy germs. Other kids do not make enough antibodies. Such children have many more illnesses than you do. They have to take many special medicines to help them stay healthy.

Some children spend their lives inside a plastic bubble. The air they breathe, their food, and their toys must all be sterilized. This is because they lack lymphocytes. Even germs that are usually harmless can make these kids very ill.

Bone marrow cells from a close relative can sometimes cure them. Special stem cells in bone marrow make millions of new germ zappers each day. In a healthy person, stem cells replace those lymphocytes that die of old age or from fighting germs.

In a child without lymphocytes, stem cells from someone else can settle in this child's bone marrow. If the stem cells start making lymphocytes, then the child can safely leave this bubble world.

Sometimes germ zappers get it wrong! They turn into destroyer cells, mounting a huge attack on substances that are of no danger to your body at all.

The sneezing, itching, and watery eyes of hay fever, and the breathing problems and wheezing of asthma, are caused by your germ zappers overreacting.

The cells that cause this problem are called mast cells and basophils. They carry chemicals that can zap human cells as well as germs.

Things like pollen, feathers, and skin cells, and saliva from cats and dogs, can cause hay fever.

Defender cells mistakenly think these are harmful. They make an antibody that sticks onto mast cells or basophils. When a tiny pollen grain meets its antibody, the mast cells and basophils explode, releasing cell-destroying chemicals that make you feel ill.

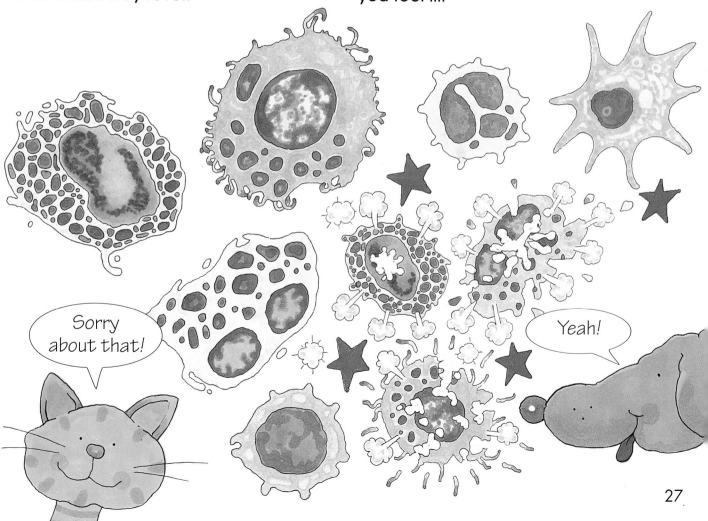

27

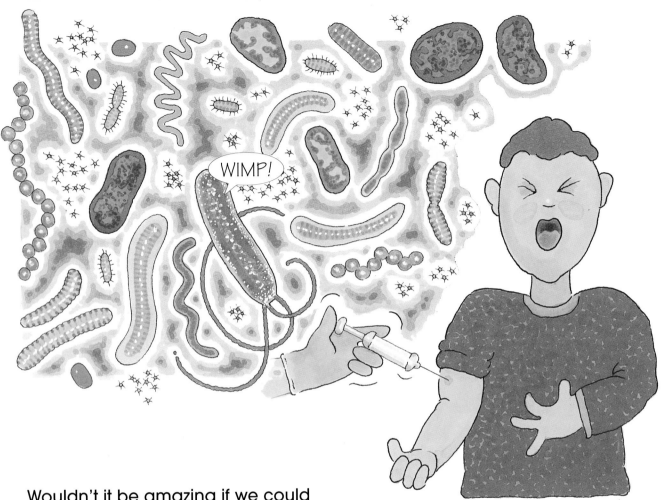

Wouldn't it be amazing if we could wipe out the dangerous bacteria and viruses that cause us so much trouble?

For some diseases, this may be possible because of a very special discovery called vaccination (vax-sin-ay-shun).

Thanks to vaccination, you and your friends lead much healthier and safer lives.

You may have already been vaccinated against some bacteria and viruses that used to harm and even kill children.

For instance, the measles virus used to make millions of kids very ill. Then scientists made a vaccine from measles virus that was carefully neutralized to do no harm. When the vaccine is given to children, their lymphocytes react as if the virus is still dangerous. The "measles squad" multiplies rapidly, making antibodies and killer cells.

If the original measles virus attacks a child after he or she has been vaccinated, the lymphocyte measles squad "remembers" the virus in double-quick time. They defeat it before any damage is done. Vaccinated children are protected from measles for the rest of their lives.

The measles virus at the top of this page has been magnified over 200,000 times.

29

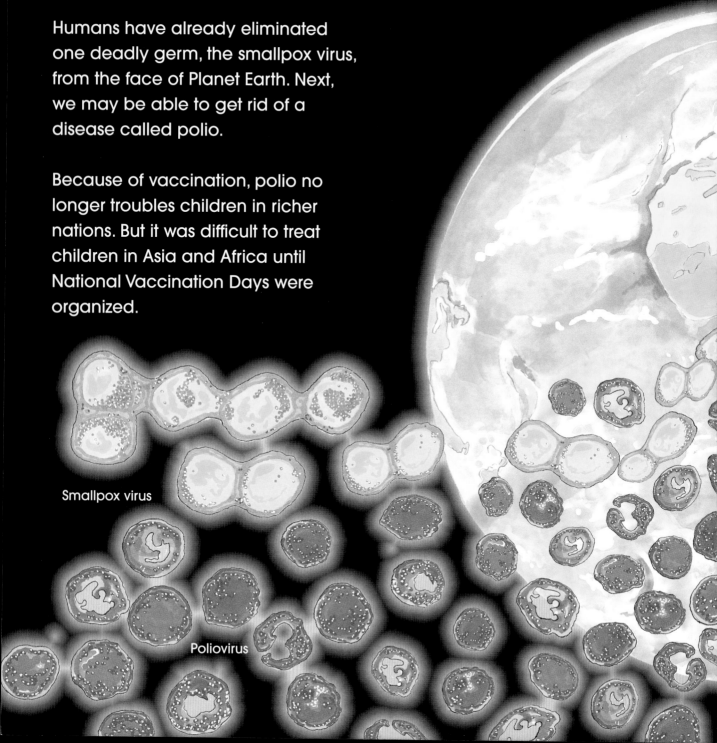

Humans have already eliminated one deadly germ, the smallpox virus, from the face of Planet Earth. Next, we may be able to get rid of a disease called polio.

Because of vaccination, polio no longer troubles children in richer nations. But it was difficult to treat children in Asia and Africa until National Vaccination Days were organized.

Smallpox virus

Poliovirus

In 2000, 550 million children in 82 different countries were protected against polio. That's almost one-tenth of the entire population of our planet!

The whole world may soon be free of the poliovirus—thanks to vaccination and our amazing—

GERM ZAPPERS!